Farmyard
Counting Book

For Hannah

First published in 1990 by
André Deutsch Limited
105–106 Great Russell Street, London WC1B 3LJ
ISBN 0 233 98468 2

Farmyard
Counting
Book

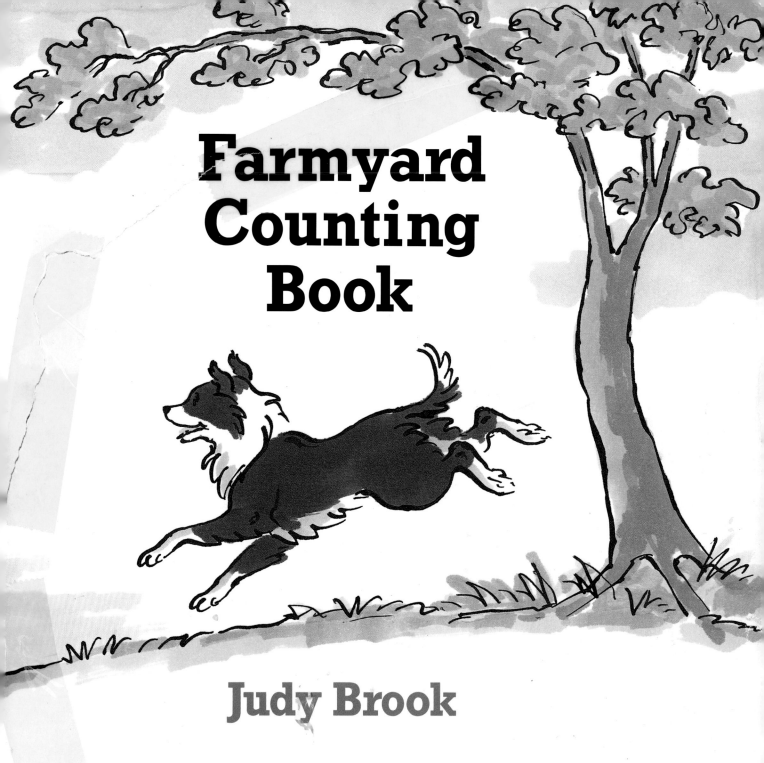

Judy Brook

ANDRE DEUTSCH

0 Nought

"Where are all the animals?" wondered Bob, the sheep dog. "There

are none in the farmyard. I'll go and see how many I can find."

By the time he had looked everywhere, he had found . . .

1 One

1 strong carthorse frisking in the field.

2 Two

2 hissing geese, who flew at Bob.

3 Three

3 jumpy cows. They didn't like Bob's barking.

4 Four

4 cross goats who warned Bob off.

5
Five

5 friendly cats on the haystack.

6 Six

6 noisy turkeys – who weren't friendly at all.

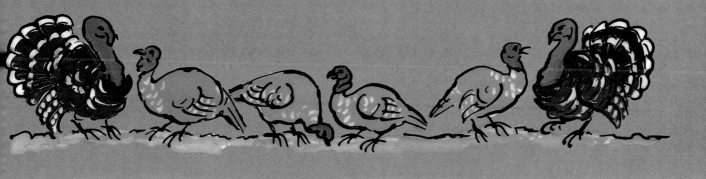

7 Seven

7 quacking ducks flying from the pond.

8
Eight

8 plump pigs. The mother pig chased
Bob away from her piglets.

9

Nine

9 fussy hens, flapping in the farmyard.

Bob had found all the animals, but he wasn't happy. He was a sheep dog

and there were no sheep on the farm.
"Poor Bob," purred the tabby cat.

Next day the farmer took Bob to market. "Will he buy some sheep for

me to look after?" Bob wondered . . .
He did.

10 Ten

"10 fine, fleecy sheep," barked Bob.

"Now I am a real sheep dog."

Bob is very happy now.
He guards the sheep carefully
and hopes there will soon
be lambs for him to look
after as well.